Flora of Malaysia
Illustrated

Flora of Malaysia
Illustrated

DAVID T. JONES

Water-colours by
PENNY GERMAN

KUALA LUMPUR
OXFORD UNIVERSITY PRESS
OXFORD SINGAPORE NEW YORK
1993

Oxford University Press

Oxford New York Toronto
Delhi Bombay Calcutta Madras Karachi
Kuala Lumpur Singapore Hong Kong Tokyo
Nairobi Dar es Salaam Cape Town
Melbourne Auckland Madrid

and associated companies in
Berlin Ibadan

Oxford is a trade mark of Oxford University Press

Published in the United States
by Oxford University Press, New York

First published 1993

British Library Cataloguing in Publication Data
Data available

Library of Congress Cataloging-in-Publication Data

Jones, David T. (David Thomas), 1952–
Flora of Malaysia illustrated/David T. Jones: illustrations by Penny German.
p. cm.
Includes bibliographical references and index.
ISBN 967 65 3044 1:
1. Botany—Malaysia. 2. Botany—Malaysia—Pictorial works.
3. Botanical illustration. 4. Watercolor painting. I. German, Penny, 1957– . II. Title.
QK365.J66 1993
581.9595—dc20
93-31124
CIP

Typeset by Typeset Gallery Sdn. Bhd., Malaysia
Printed by Kyodo Printing Co. (S) Pte. Ltd., Singapore
Published by Oxford University Press,
19–25, Jalan Kuchai Lama, 58200 Kuala Lumpur, Malaysia

Dedicated to the memory of
Kabir and Arjun Talwar

God gave us memories so that
we might have roses in December.

James Barrie

Foreword

IN this era of growing environmental awareness, those tropical countries fortunate enough to have a heritage of scientific information that provides, at least broadly, a summary of their natural resources, can begin or continue the tradition of bringing this information into forms that provide both educational and aesthetic benefits, such as the present book.

Tropical countries are biologically rich, but they are at risk through environmental degradation, overexploitation, and pollution. They are also complex, and for the most part insufficiently explored and studied. Scientists—botanists, ecologists, zoologists, and the like know a good deal about them, but all are agreed that they know far too little. The only way to learn more is to have the opportunity to study at first hand the great tropical forests and their inhabitants. But these have been so decimated in the last few decades, and are disappearing so rapidly today, that it is vitally important that conservation measures of all kinds are put into action.

In the pursuit of conservation, however, we must not forget the numerous values and virtues of native flora (and fauna), not the least among them being their capacity to intrigue us by their variety of form and colour, and to please us by their beauty and intricacy. Lists of scientific names (or even of common names) are hardly sufficient to attract laymen, students, and tourists (unless of course they already know and can visualize what these names refer to).

Guides, handbooks, and reference works of all kinds have used illustrations for various purposes. Of all the means of representing objects and creatures, nothing compares to an illustration for immediacy. Beyond this, however, is the possibility of enjoyment created by artistic finesse. Why should the education packet not be as attractive as it is nutritious to the mind?

In this work, the author, David Jones, and the illustrator, Penny German, have joined forces to provide both ingredients. Jones, with his long experience and extensive fieldwork in Malaysia and the surrounding area, has the knowledge of the flora to focus on the things the reader will want to know and will enjoy learning. As botanical author, he provides the firm basis for nomenclatural accuracy, indicating the correct scientific (and for most, vernacular) name of each species, as well as the family to which it belongs. German, with her skilfully drawn and attractively coloured illustrations, brings the plants to life—accurately and realistically. The result is a pleasing combination of the scientific and the artistic that adds knowledge to gracefulness and precision to communication.

Choosing a selection of plants to illustrate for a volume of reasonable size is a difficult task. One needs to consider whether common or rare plants should be included; big, colourful ones, as against the usually much more abundant shy, small, and sometimes rather drab ones; endemic or common species; garden plants or plants of remote forests and mountain tops; useful plants or scientifically fascinating but commercially unexploited ones.

There is obviously no one right answer to this kaleidoscopic conundrum. In solving it, the author and illustrator have developed a simple hierarchy of plant groups, each represented by 3–5 species. The groups are not botanical taxa, nor yet hyperlogical groups, but they furnish a human context: climbers, water plants, mangrove species, wayside plants, flowering trees, garden shrubs, herbs, sea-shore plants, conifers and their relatives, mountain species, orchids, fruit trees, gingers, pitcher plants, and ferns and their allies. This pleasantly eclectic arrangement is convenient and makes sense enough to both the tyro and the professional. In each group, the plants are beautifully illustrated, identified, and discussed in a knowledgeable, informative, and friendly manner.

A book like this will surely find a niche in the new tourist's pocket as well as the younger student's book bag; but it will also show up on the bookshelf in the library, and on the desk of the executive and the scientist, who appreciate either the easy morsels of knowledge, or the pleasant and curious aesthetic aspects—or both.

Malaysia is fortunate in having a number of very good books about her native plants. Classics such as E. J. H. Corner's *Wayside Trees of Malaya*, R. E. Holttum's *A Revised Flora of Malaya*, and M. R. Henderson's *Malayan Wild Flowers* each provide an entry into the marvellous, and increasingly rare, world of Malaysian botany, some through the sterner routes of science, some through less steep if more circuitous approaches. Jones and German (as this book will no doubt come to be referred to) have built a charming pathway, passing many fascinating and representative species, into the Garden of Knowledge (Rimba Ilmu): a small version of the vital, diverse Malaysian forests. This is appropriate, for Jones administered for several years the 'real' Rimba Ilmu—the University of Malaya Botanical Gardens. Readers of this work may be encouraged to visit these and other botanical gardens, and then to enter the finest gardens of all—the magnificent, pristine tropical forests in Malaysia or elsewhere.

If knowledge provides pleasure, so let pleasure—enjoying this book—provide knowledge; just as Rimba Ilmu may be rearranged to Ilmu Rimba, the Knowledge of Gardens.

BENJAMIN C. STONE
Reader in Botany, Emeritus, University of Malaya
Senior Research Botanist, Botanical Research Institute of Texas, Fort Worth

Preface

IN *The Complete Flower Paintings and Drawings of Graham Stuart Thomas*, Graham Stuart Thomas writes, 'It is a notable fact ... while there has been no dearth of artists and writers in horticulture, very seldom have the two talents occurred in the same individual.' Thus arose this collaborative effort to bring words and art together in a celebration of the beauty of plants and flowers. Throughout history, artists and writers have tried to capture the essence and grace of flowers. From the floral motifs used by the ancient Egyptians to the rather mystical reverence given flowers in the art and literature of the Far East, the lure of flowers has not changed, and so it is that we, too, have found the fresh and hopeful quality of flowers compelling.

While the title of the present volume, *Flora of Malaysia Illustrated*, may evoke in the reader's mind an image of the Malaysian flora in all its complexity and grandeur, this book is not a 'Flora' in the sense of providing a comprehensive treatment of plant life found in Malaysia. That approach has already been followed by far more ambitious and competent writers, resulting in the publication of a wide range of popular and technical works covering most of the major groups of Malaysian plants.

Our *Flora* is a portfolio of plant portraits selected from among the thousands of species which live in Malaysia, whether they be wild forest dwellers or cultivated garden ornamentals. The 58 plant species which finally made their way into this volume were chosen because they possessed innate beauty in form, texture, and colour, and because they displayed diversity in growth habit and habitat. Although a preliminary list of candidate species was drawn up at the outset, it was continually reviewed and amended, for as we quickly discovered, the ephemeral nature of flowers—some last for just a few hours or days—can thwart attempts to 'capture' them. In some cases, plants were chosen because they were easily available, accounting for the inclusion of a number of fairly common plants. In others, it was sheer determination (as with the Peacock Begonia), and even pure luck (as with Skeat's Bulbophyllum) that enabled us to locate the plant and include it in the present work.

The text of the book is composed of a brief Introduction, which serves to place the entire Malaysian flora in a biological and historical context, and descriptions of the individual plants. The plant descriptions are botanical vignettes designed to provide information of general interest about the plant's nativity and distribution in Malaysia, appearance and growth habit, traditional uses, botanical and vernacular names, and classification. Some readers will find gaps in the information, an unavoidable consequence of trying to keep the text brief, and we apologize for any shortcomings arising from this. Those wishing to know more about the plants treated here, or about any other aspect of the Malaysian flora, are referred to the annotated Bibliography provided.

The real value of this book, we hope, will develop out of an appreciation for the verity of the botanical portraits, and for their aesthetic content. The limitations imposed by the requirements of the writer and the format of the book have not diminished the artistic quality of the original water-colour illustrations. The enthusiasm for tropical plants has become widespread in recent years and some of the plants portrayed here will be instantly recognized. What should be strikingly clear is the way in which even the most mundane plant (take, for example, the common weed, *Trimezia martinicensis*, the Forenoon Yellow Flag) can appeal to our artistic and intellectual senses.

With this book, we acknowledge our joy in the beauty of plants and flowers. If it elicits delight, then the book will have fulfilled its purpose. And if it moves you to explore the world of plants beyond its pages, may it help to guide you down the path of discovery.

May 1993

DAVID T. JONES
PENNY GERMAN

Acknowledgements

THE authors are especially grateful to the staff of the University of Malaya's Department of Botany and Botanical Gardens (Rimba Ilmu) for providing support, assistance, and materials during the preparation of this book. We would also like to thank the staff of the Forest Research Institute Malaysia, the Orchid Garden in Kuala Lumpur, the Singapore Botanic Gardens, and Kinabalu Park, Sabah, for their co-operation in allowing us access to certain plants in their living collections, and the Royal Botanic Gardens, Kew, England, for their assistance with plant identification. We are also indebted to the following individuals for their generous assistance: Helmi Baharuddin, Henry Barlow, M. Bresin, Norman and Mary Cooper, R. German, Noorma Wati Haron, Halijah Ibrahim, Lynette Carrington-Kerslake, Sandra Leininger, Ilsa Noor, George Ong, Lily Ramli, Lina Santiago and family, and E. Soepadmo.

For permission to reproduce illustrations for use in this publication, we would like to thank the following: Mr and Mrs T. Cameron (*Ipomoea acuminata*), Cdr. S. Pegg (*Hibiscus rosa-sinensis*), Mr and Mrs R. Hudson (*Etlingera elatior*), Mr F. Davies (*Nepenthes fusca*), and Sir Nicholas and Lady Margaret Spreckley (*Nepenthes sanguinea*).

Contents

Plates

Introduction

Situated upon the Equator, and bathed by the tepid water of the great tropical oceans, this region enjoys a climate more uniformly hot and moist than almost any other part of the globe, and teems with natural productions which are elsewhere unknown. The richest of fruits and the most precious of spices are here indigenous. It produces the giant flowers of the Rafflesia, the great green-winged Ornithoptera (princes among the butterfly tribes), the man-like Orang-Utan, and the gorgeous Birds of Paradise.

(Alfred Russel Wallace, 1869)

WITH these words opens the historical narrative, *The Malay Archipelago*, written by the eminent English naturalist and correspondent of Charles Darwin, which documents the author's 8-year sojourn through this region over 130 years ago. Like so many other visitors to the Archipelago, before and since, Wallace was moved to write about the lure of these lands—the profusion of their plants and animals, the diversity of their native peoples—and the fascination they held for him. More than a century later, the natural beauty of the region continues to enthral and inspire, no part more profoundly than the magnificent rain forests.

The Archipelago described by Wallace represents only a portion of tropical Asia, that region extending from the plains of western India eastward to Indo-China and across the scattered islands of Malesia. The rain forests here, reduced to fragments from a once continuous swath which covered the region, are still resplendent in their diversity of life forms. These forests—the Indo-Malesian tropical rain forests—are considered to be among the richest in the world in terms of plant life. Floristic studies have shown that small areas of them are richer in tree species than equivalent areas in both the African and American equatorial rain forests. In Malesia alone, that part of South-East Asia comprising the political entities of Malaysia, Indonesia, Papua New Guinea, and the Philippines, there are an estimated 20,000–30,000 species of flowering plants representing approximately 10 per cent of the world's total. Of this, 40 per cent of the genera and more than half of the species are believed to be endemic, which means that they occur here and nowhere else.

There are several reasons for this prodigious variety of plant life. Despite a near uniform climate of high temperature and moisture which exists across the region, local hydrologic, soil, and topographic conditions combine to create a wide range of microenvironments which support an immense variety of plants. Most strikingly, the complex structure of the tropical forest itself creates a whole series of habitats by providing a framework as well as an internal microclimate for the evolution and coexistence of many plant species, some of which are highly specialized in the ecological roles they fill.

The relative absence of disruptive palaeoclimatic, and to a certain extent, geologic events, in some parts of Malesia, has also contributed to the floristic richness of its rain forests. Despite past fluctuations in climate and sea levels, and episodes of major volcanic activity and mountain building, affecting migration and overall distribution patterns of both plants and animals in the region, the vegetation of some areas has persisted more or less in its present state for millions of years. Such stability has allowed species to evolve uninterrupted and accumulate over a long period of time.

Within this region of great biological diversity lies Malaysia, comprising Peninsular Malaysia (West Malaysia) and the states of Sabah and Sarawak (East Malaysia) in northern Borneo. It is estimated that between 9,000 and 10,000 species of angiosperms (flowering plants), gymnosperms, ferns, and bryophytes are indigenous to Peninsular Malaysia, 90 per cent of which are angiosperms. The strategic position of the peninsula, linking continental Asia to the north with the rest of Malesia to the south, has served as an important migratory track between these two great floristic regions, the resultant merging of the Asiatic and Malesian elements contributing significantly to the botanical wealth of the peninsula.

Sabah and Sarawak are situated approximately 1 000 kilometres to the east of Peninsular Malaysia, and at the same latitude. Their combined flora is comparable in both size and composition to that which exists in the peninsula, and it is similarly characterized by a high number of endemic genera and species. By virtue of the fact that Sabah and Sarawak lie near the geographical centre of the Malesian region, their flora, like that of Peninsular Malaysia's, contain elements of Asiatic, Malesian, and, to a lesser extent, Australasian origin. Botanists have, in fact, determined the western Malesian region to encompass a singular phytogeographical unit based on the close floristic similarities between Borneo, Sumatra, and Peninsular Malaysia.

What appears to some to be a uniform expanse of green vegetation blanketing the Malaysian landscape is, in fact, a mosaic consisting of natural forest types, managed agricultural lands, and disturbed marginal habitats. The greatest wealth of plant life is found in the rain forests themselves where the dynamic interplay of evolutionary and ecological processes has produced a variety of forest formations made distinct by their unique structures and constituent life forms—trees, shrubs, herbs, climbers, epiphytes, parasites, among others—embracing a wide range of different plant families, genera, and species.

The rain forest formations are extremely varied, and reveal immense range in habitat and vegetation. On the flat coastal plains, there are extensive wetlands dominated by mangrove and brackish-water swamp forests, with a complex zonation of distinct plant and animal communities. Inland, peat swamp forests which receive their water from rainfall only, and freshwater swamp forests fed by water from lakes and streams, are found, the latter also developing in river valleys. Beach forest, the narrow strip of vegetation which forms on dry, sandy, or rocky sea-shores, including islands, prevails in some coastal areas.

The lowland evergreen tropical rain forest is the most extensive and luxuriant forest type found in Malaysia. Despite the heterogeneity of its flora, the general appearance of the forest in different areas is rather uniform and typified by the presence of tall buttressed trees (predominantly dipterocarps), large woody climbers (lianas), and sparse ground vegetation. Where gaps occur in the forest canopy, pockets of secondary forest develop, characterized by fewer, different species of plants. In areas where the soil is derived from an underlying bedrock of limestone, ultrabasic rock, or siliceous material, the lowland forest is replaced by still different forest types of contrasting structure and floristic composition where endemic species are not uncommon.

The cooler temperatures coinciding with increased elevation mark the transition to the montane rain forest. This formation extends from the foothills to the highest mountain peaks, and is marked by the low stature of the vegetation and the large number of epiphytic ferns and mosses. The floristic composition of these forests is distinguished by the high representation of plant families and genera having affinities with the temperate floras of continental Asia and Australia.

The diverse organisms inhabiting Malaysia's rain forests interact with the environment, and with each other, to form a complex and stable ecosystem. The limited information available on plant–animal interactions suggests that they cannot be ignored as a factor in explaining the rich plant life found in these forests. Rain forest animals, especially insects, birds, bats, and small mammals, acting as pollinators of flowers or as dispersal agents for fruits and seeds, have undoubtedly stimulated the formation of plant species through long-term coevolutionary relationships with them. Interactions between ants and certain plants in which they nest (myrmecophytes) further illustrate how close associations with animals can afford plants with increased survival value, advantageous in the highly competitive rain forest environment.

Like so many other tropical areas throughout the world, the Malaysian flora has been strongly influenced by the activities of Man. In many ways, it has become enriched through the deliberate or inadvertent introduction of exotic species, many of which are now found everywhere in the Tropics. It has been argued that, although imposing, these 'cosmopolitan' elements are in danger of replacing the native flora as they encroach upon it, and give it a rather artificial appearance. But even more threatening is the selective removal of indigenous plant species which impoverishes the rain forest ecosystem, and large-scale deforestation which endangers its very existence. Tropical rain forests require close management if they are to continue performing their protective and regulative functions in the environment, and their productive one, namely providing many important food products and useful raw materials, ranging from timber and minor forest products to drugs and organic chemicals.

The unprecedented pace of development in Malaysia during the twentieth century has placed enormous pressure on the country's rain forests. As they diminish in size, the threat of plant extinctions and gradual floristic impoverishment becomes more certain. Whatever is attempted or done, a time will eventually arrive when, apart from forest reserves, national parks, and botanical gardens, Malaysia's tropical world will be irreparably exhausted of its once munificent botanical treasures.

Flora of Malaysia Described and Illustrated

Climbing Plants

FEW tropical plants can compare with the versatility of climbing plants which use various means—hooks, tendrils, and twining stems—to assist in their struggle upwards. Malaysia's forests are replete with a multitude of native climbing plants, from delicate, herbaceous vines to large, woody lianas, some even successfully introduced into cultivation and embellishing many a home garden. Still others represent the myriad climbers of exotic origin which are now a regular feature of the Malaysian scene.

1 *Aeschynanthus longicalyx* Ridley

Lipstick Plant

THIS beautiful climbing epiphyte is readily known by the conspicuous bright red flowers arising from the leaf axils, or clustered near or at the ends of slender drooping stems. The fleshy smooth-edged leaves are sometimes tinged purple. The two-lipped tubular corolla is embraced by the large toothed calyx (hence the epithet *longicalyx* from the Latin *longus*, 'long', and *calyx*, 'outer envelope of a flower'). Two pairs of stamens, fused at their anthers, protrude from the corolla mouth. The fruit is a long and slender capsule which splits longitudinally to release the numerous minute seeds.

The Lipstick Plant is indigenous to Peninsular Malaysia where it is not infrequent in montane forest along the Main Range. It is one of over a dozen species of the genus found in Malaysia, all members of the predominantly tropical plant family *GESNERIACEAE*, recognized throughout the world for its many cultivated ornamentals. *Aeschynanthus* is especially well-suited for suspended containers which display the elegance of its flowers and foliage most effectively. Its name derives from the Greek *aischyne*, 'to be ashamed', and *anthos*, 'flower', referring to the red flowers.

Painted from a specimen in forest on Gunung Brinchang, Cameron Highlands, Pahang.

2 *Tristellateia australasiae* A. Rich.

Galphimia Vine

INDIGENOUS to coastal areas and tidal swamps ranging from Peninsular Malaysia and Thailand eastward to Australia, the Galphimia Vine is now cultivated extensively as an ornamental plant in Malaysian gardens and parks. It is a climbing shrub with thin, woody, twining branches which bear dull green leaves. The striking inflorescences of five-petalled yellow flowers are produced all year round, often in great profusion. The stamens of each flower are initially yellow in colour and become red with maturation. Curious star-shaped fruits, composed of three united samaras with 5–8 wings each, are easily seen on flowerless stalks.

The name *Tristellateia* is derived from the Latin *tres*, 'three', and *stella*, 'star', alluding to the fruit clusters. The genus is one of three others, classified under the family *MALPIGHIACEAE*, native to Peninsular Malaysia. The majority of the Malpighiaceae are small trees or shrubs from warm and tropical regions of Central and South America. The vernacular name, Galphimia Vine, is a reflection of its resemblance to *Galphimia*, a related genus of shrubs and small trees native to tropical America.

Painted from a specimen on the campus of the University of Malaya, Kuala Lumpur.

3 *Quisqualis indica* L.

Rangoon Creeper, Akar Dani

THIS large woody climber, a native of South-East Asia, is distributed from Burma eastward to Papua New Guinea, including Malaysia and the Philippines. The Rangoon Creeper grows naturally along river banks, and has long been cultivated in the Tropics and subtropics (it was introduced to Europe more than a century ago) because of its horticultural value. The fragrant flowers are borne prolifically in terminal bunches. The beauty of the flowers lies in the unusually long pendant calyx tube which is further enhanced by the spreading, variously coloured petals, which change from white in the newly opened blooms to reddish-pink.

The name *Quisqualis* derives from the Latin *quis*, 'who?', and *qualis*, 'what?', believed to refer to the astonishment of an early botanist at the changing colours of the flowers. The leaves, fruits, and roots of the plant have been used in traditional medicine, while the seeds (said to taste like coconut) and young shoots are eaten. The Rangoon Creeper, one of the few species of its genus which is cultivated to any great extent, is a member of the family *COMBRETACEAE*, comprising mostly tropical and subtropical species, especially African.

Painted from a specimen in a garden in Kuala Lumpur.

4 *Ipomoea acuminata* (Vahl) Roem. & Schult.

Blue Dawn Flower

A vigorous perennial vine of tropical American origin, the Blue Dawn Flower is widely cultivated throughout the tropical and warm temperate regions of the world. In Malaysia, the most splendid specimens are seen in the mountains, adorning trellises and fences around homes and gardens, and not uncommonly growing semi-wild. This shrubby climber, with stems attaining heights over 10 metres, produces a variety of leaf forms, from heart-shaped to three-lobed, with pointed tips (hence the epithet *acuminata* from the Latin, meaning 'sharpened'). From the axils of the leaves arise clusters of bright blue to violet flowers with a trumpet-shaped corolla, the limb of which can measure up to 12 centimetres in diameter. The flowers are often tinged pink, especially in bud and along the corolla tube and the five plaits of the limb in opened flowers.

The Blue Dawn Flower, so named because its flowers open early in the day, belongs to the cosmopolitan family *CONVOLVULACEAE*, the Morning Glory or Kangkong Family. The genus *Ipomoea* is recognized as an important source of food (leaves, fruits, tubers), ornamentals, hallucinogenic drugs, and traditional medicines. Its name derives from the Greek *ips*, 'a kind of worm', and *homoios*, 'resembling', in reference to the twining stems.

Painted from a specimen at Fraser's Hill, Pahang.

Aquatic Plants

WITH the exception of marine algae and the angiospermous sea grasses which inhabit shallow coastal waters, Malaysia's aquatic flora is represented by freshwater plants, the majority of which are flowering. The considerable number of native and introduced species, and the great variety of families they represent, is paralleled by their diverse growth habits—completely and partially submerged and free-floating types—and peculiar structural and reproductive adaptations, required for living in this unique environment.

5 *Limnocharis flava* (L.) Buchenau

Yellow Burhead, Jinjir

ROOTED in the mud of irrigation ditches and rice-fields throughout Malaysia, this partially submerged perennial marsh herb was introduced from tropical South America many years ago. The name *Limnocharis* derives from the Greek *limne*, 'marsh', and *chairo*, 'to delight in', and refers to the habitat. The velvety tufted leaves, rising to a height of over 50 centimetres above the water, are soft and fleshy, their angled petioles constructed of a network of air channels. The inflorescence stalk bears a terminal cluster of yellow flowers (hence the epithet *flava* from the Latin *flavus*, 'yellow'), which open early on bright mornings and close shortly after. The inflorescences and young leaves are reported to be eaten by the Javanese. The fruit, a follicetum, is divided into several sections, each containing numerous seeds. When ripe, the inflorescence stalk bends to the water surface to shed the seeds.

The Yellow Burhead is the only representative in Malaysia of the *BUTOMACEAE*, a primitive family of monocots. Because of certain leaf and floral characteristics, and the presence of white latex in its tissues, some botanists classify this species in a separate family, the *LIMNOCHARITACEAE*.

Painted from a specimen in a flood control ditch near Kelang, Selangor.

6 *Nelumbo nucifera* Gaertner

Sacred Lotus, Teratai

This Asiatic water plant hardly needs an introduction, its exquisite flower being familiar to almost everyone. The flowers are sacred to Buddhists, and the lotus motif has been used extensively in art forms throughout Asia. It was taken to Egypt by Man about 500 years BC, where it is also venerated as a sacred plant. The large green bowl-like leaves are raised high above the water on stout petioles, over 1 metre long. The solitary flowers, typically pink (white, red, and striped varieties also exist), with numerous stamens and funnel-shaped 'seed pods', are also produced on long stalks which can overtop the leaves. The seed-like fruits, embedded in the curious flattened receptacle, are remarkable in being able to germinate after being deposited in river mud for several hundred years.

Cultivated widely for its edible plant parts—receptacle, fruits, rhizome, and petiole—the Sacred Lotus is grown less for food than for the beauty of its flowers. The Malays have reportedly used the petals to treat syphilis. This species is classified under the family *NYMPHAEACEAE*, together with the Water Lily, *Nymphaea*, or in its own family, the *NELUMBONACEAE*. The name *Nelumbo* is derived from the Sri Lankan one for this species.

Painted from a specimen at Tasek Cini, Pahang.

7 *Monochoria hastata* (L.) Solms-Laub.

Pondweed

A native of Malaysia and other parts of the Indo-Malesian region, the Pondweed is a freshwater perennial herb colonizing ditches, rice-fields, and other wet, swampy habitats. The broad spear-shaped leaf blades (hence the epithet *hastata* from the Latin *hasta*, 'spear') stand on slender stalks nearly 1 metre long. Emerging from a leaf-like sheath on the leaf stalk, the inflorescence bears a cluster of crowded flowers with blue to light violet petal-like perianth lobes and yellow anthers. The name *Monochoria* is derived from the Greek *monos*, 'single', and *choris*, 'apart', referring to the presence of a single large stamen which stands apart from the others. The flowers open a few at a time, the uppermost first, and last only a day. The fruits are small round capsules that can easily be seen rising above the encircling flowers. The inflorescences and leaves of this plant are eaten as a vegetable, while the roots are said to be used medicinally in Sulawesi.

The Pondweed is a member of the family *PONTEDERIACEAE*, to which the Water Hyacinth or Keladi Bunting (*Eichhornia crassipes* Solms-Laub.) belongs, a noxious aquatic weed introduced from South America at the turn of the century.

Painted from a specimen in the Kelang Valley, Selangor.

Mangrove Plants

To the unaccustomed eye, the profusion of stilt roots, flying buttresses, and breathing roots which characterize mangrove plants is an astonishing sight. These structures represent but a few of the myriad adaptations found in the salt-loving mangrove vegetation of tropical and subtropical tidal flats, estuaries, and lagoons. Although comparatively poor floristically, mangroves comprise species which are widely distributed geographically, on account of their fruits having adapted for dissemination by sea currents.

8 *Merope angulata* (Willd.) Swingle

Mangrove Lime, Limau Lelang

This curious shrub grows along the banks of streams which pervade the landward side of tidal swamps ranging from the mouth of the Ganges to Peninsular Malaysia, Borneo, and Papua New Guinea. From the axils of the dark green leathery leaves arise both formidable spines, up to 5 centimetres long, and fragrant delicate flowers with bright white petals. The local vernacular name, Lelang, is believed to derive from the Malay word *helang*, 'eagle', alluding to the talon-like spines. The peculiar fruits of the Mangrove Lime are buoyant three- to four-sided berries (hence the epithet *angulata* from the Latin *angulus*, 'corner') with thick aromatic skins dotted with oil glands. They are filled with large flattened seeds embedded in a sticky viscous fluid. Schoolchildren in the peninsula have been observed applying this transparent mucilage to their fingernails as a sort of polish. The fruits are said to be used medicinally for the treatment of colic and other intestinal complaints.

The Mangrove Lime, a highly specialized plant, and the sole member of its genus, is allied with *Citrus* in the family *RUTACEAE* (see also *Murraya paniculata*). The name *Merope* derives from Merope of Greek mythology, one of the Heliades.

Painted from a specimen in mangrove forest near Pengkalan Kempas, Negeri Sembilan.

9 *Sonneratia alba* J. J. Sm.

Perepat

PEREPAT, a small to medium sized tree found throughout the tidal forests and sheltered estuarine mud-flats of the Old World Tropics, is Malaysia's most common mangrove species. It is an easy tree to recognize because of its thick, fleshy, orbicular leaves, and large, white, powder puff-like flowers (hence the epithet *alba* from the Latin *albus*, 'white') found in terminal clusters. A related Malaysian species, *Sonneratia caseolaris* (L.) Engl., differs in having dark red petals and stamens. The flowers open at dusk, and the pink-tinged petals and stamens drop before dawn. The fruit is a buoyant, globose, green or yellowish berry, tipped by the slender style, which sits on the persistent star-like calyx. Pollination of the flowers and dispersal of the fruits and seeds is probably aided by bats which feed on these structures.

The genus *Sonneratia*, named in honour of P. Sonnerat, the eighteenth-century French explorer and botanist, belongs to the Old World family *SONNERATIACAEAE*. Certain Malaysian species of the family are a source of firewood, building material, and traditional medicines. The ripe fruits of Perepat are edible; the leaves are eaten, raw or cooked, as a vegetable.

Painted from a specimen on the coast near Sementa, Selangor.

10 *Bruguiera gymnorrhiza* (L.) Lam.

Tumu Merah

NAMED in honour of J. G. Bruguieres, a contemporary of the famous eighteenth-century French naturalist Jean Baptiste Lamarck, the genus *Bruguiera* is distributed from tropical East Africa to Australia and the Pacific. Tumu Merah is one of five species of the genus found in Malaysia, where it prefers the drier, well-aerated soils of mangrove forests. The epithet *gymnorrhiza* derives from the Greek *gymnos*, 'bare', and *rhiza*, 'root', and refers to the exposed breathing roots (pneumatophores) found in this species. The leaves of this medium sized tree are large, elliptic, and leathery. The flowers arise singly or in groups of 2–5 from the leaf axils, and bear stiff, linear, pubescent petals which are at first white, then change to brown. The most outstanding feature of the plant is the bronze-coloured, cigar-shaped radicle, or primary root, which develops from the fruit while still attached to the tree, serving to anchor the developing seedling into the mud when it drops.

Tumu Merah has provided many useful products of local importance: charcoal, building material, tannins and dyes, food, and local medicines. It is classified under the *RHIZOPHORACEAE*, a predominantly Old World family including both mangrove and inland species.

Painted from a specimen on the Selangor coast.

Wayside Plants

THERE is no better introduction to the Malaysian flora than through a study of its wayside vegetation—common plants that live along the margins of forests and roadsides and in waste places and secondary vegetation (*belukar*)—where the diversity of species and profusion of growth forms underlie its often chaotic appearance. The ability of these plants to colonize disturbed ground and spread rapidly accounts for their wide distribution, and enables them to cover large areas. Some species introduced from abroad have become well-established in these habitats.

11 *Calotropis gigantea* (L.) Aiton f.

Giant Milkweed, Remigu

THE Giant Milkweed is indigenous to India from where it spread to various areas in southern Asia. In Malaysia, it is frequently found around villages where it is sometimes cultivated, and in dry areas, especially near the sea. This freely branched shrub, growing to heights of up to 4 metres, produces stout stems which bear large fleshy leaves with prominent veins on their undersides. The complicated flowers, formed in clusters in the axils of the leaves, consist of a wheel-shaped corolla with a crown (corona) of fleshy appendages joined to a central column. The name *Calotropis* is derived from the Greek *kalos*, 'beautiful', and *tropis*, 'keel', alluding to the flowers. They are made into leis in Hawaii. The fruit, a fleshy inflated follicle, is filled with numerous flattened seeds, each with a plume of fine hairs. The stems of the Giant Milkweed yield high quality Yercum fibre, and the leaves, which produce a white milky latex, have entered into traditional remedies for rheumatism and skin disorders.

This species is a member of the family *ASCLEPIADACEAE*, largely represented in Malaysia by wild climbing plants such as the beautiful Wax Plant (*Hoya*) and the ant-inhabited *Dischidia*.

Painted from a specimen near Kuantan, Pahang.

12 *Melastoma malabathricum* L.

Singapore Rhododendron, Sendudok

THIS variable plant, distributed from the Mascarene Islands eastward to the Pacific, is a common element of regrowth vegetation in the lowlands and hills of Malaysia, especially along forest margins and in open clearings. Shrubs are everflowering, their attractive flowers produced in terminal clusters, with colours ranging from pink to violet-mauve and enhanced by the yellow-stalked stamens, each with a conspicuous appendage. The scaly berry-like fruits are filled with an edible, purple, pulpy mass containing tiny seeds. The name *Melastoma* is derived from the Greek *melas*, 'black', and *stoma*, 'mouth', in reference to the staining of the mouth if the fruits are eaten. These sweet fruits are, in fact, relished by birds and small mammals which assist in the wide dispersal of the seeds. The leaves and roots of the Singapore Rhododendron have been employed in traditional treatments for a variety of ailments, including dysentery, ulcers, haemorrhoids, and flatulence.

The plant is classified under the *MELASTOMATACEAE*, a family well-represented in the Malaysian flora by herbs, shrubs, and trees, many with showy and very beautiful flowers.

Painted from a specimen in secondary forest at Kuala Kubu Baru, Selangor.

13 *Dillenia suffruticosa* (Griff.) Martelli

Simpoh Air

THE genus *Dillenia*, commemorating J. J. Dillenius, the German botanist of the early eighteenth century, comprises 60 species distributed naturally from Madagascar eastward to Australasia. One of the commonest species in Malaysia is the everflowering Simpoh Air which forms dense thickets in secondary forests and swamps in the lowlands and hills. The epithet *suffruticosa* is derived from the Latin *sub*, 'under', and *fruticosus*, 'shrubby', and alludes to the bushy growth habit of the plant. The large cabbage-like toothed leaves and broad yellow-petalled flowers readily distinguish this plant from other *belukar* vegetation. The star-like pink 'fruits' (actually an aggregation of dehisced follicles) contain seeds enveloped by a fleshy scarlet-coloured aril. This is particularly attractive to birds which help to disperse the seeds across the landscape.

While no economic uses for Simpoh Air have been reported, several other Malaysian species of *Dillenia*, such as Simpoh Gajah (*D. reticulata* King), can attain timber size, and are valued for their attractive wood. All Simpoh belong to the *DILLENIACEAE*, a small family of about 200 species found in the Old and New World Tropics.

Painted from a specimen in secondary forest near Kuala Lumpur.

14 *Cassia alata* L.

Candlestick Cassia, Gelenggang

As suggested by its vernacular name, the Candlestick Cassia is a striking shrub which produces erect spikes over 0.5 metres in length, bearing crowded bright yellow flowers and orange bracts. It is indigenous to tropical America, but is naturalized and even cultivated in other tropical parts of the world. In Malaysia, the Candlestick Cassia is frequently encountered around villages and in exposed waste places, where it grows to a height of 2–3 metres. The somewhat straggly shrub bears pinnate leaves with 8–20 pairs of stiff leaflets and narrow winged pods (hence the epithet *alata* from the Latin *alatus*, 'winged'), containing numerous flattened seeds. The leaves and pods of this species are valued by Malays and Chinese for treating ringworm and other skin afflictions, justifying the common name Ringworm Cassia. An infusion of any part of the plant is said to possess laxative properties.

Several other *Cassia* species are cultivated ornamentals in Malaysia, the finest being the Golden Shower, *C. fistula* L., which also happens to be cultivated in certain parts of the world for its medicinal pods and leaves. The Cassias are members of the *LEGUMINOSAE* (see also *Bauhinia purpurea*), the third largest flowering plant family in the world.

Painted from a specimen along the old Gombak–Kuantan Road near Kuala Lumpur.

Flowering Trees

TREES are the dominant life form in tropical forests, determining their overall structure and affecting their functioning. The enormous diversity found in the size, branching patterns, foliage, flowers, fruits, and seeds of Malaysian trees is paralleled by the large number of families, genera, and species that occur naturally here, greater than in any other tropical region. When brought into cultivation, the flowering trees of both indigenous and exotic origin provide the most impressive displays of floral form and colour.

15 *Gardenia carinata* Wallich

Kedah Gardenia, Randa

THIS relative of the popular garden shrub, the Cape Jasmine (*Gardenia jasminoides* Ellis) from China, is a small forest tree from the hills and lowlands of Peninsular Malaysia and Thailand. With its large leaves and richly coloured and exceedingly fragrant flowers, the Kedah Gardenia ranks among the most beautiful of Malaysia's native flowering trees. The newly formed flowers are cream-coloured, opening at night to attract moths which act as pollinating agents, before they change to yellow and finally orange. Despite its striking beauty, this species was largely ignored as an ornamental tree, except in the north. It can now be seen widely planted in gardens and parks, and along avenues. The fruit is a berry with a hard woody wall which can be cracked between the teeth to extract the pulpy mass of seeds.

The genus *Gardenia*, commemorating Alexander Garden, the eighteenth-century English naturalist, comprises about 60 species of shrubs and small trees native to the Old World Tropics. All belong to the *RUBIACEAE* (see also *Ixora javanica*), the fifth largest family of flowering plants in the world, and the second largest in Malaysia.

Painted from a specimen on the campus of the University of Malaya, Kuala Lumpur.

16 *Mesua ferrea* L.

Ironwood, Penaga Lilin

LIKE other species of the genus *Mesua* (commemorating Johannes Mesue, the nineteenth-century physician from Damascus), the Ironwood has extremely hard wood, hence the epithet *ferrea* from the Latin *ferrum*, 'iron'. In the late nineteenth century, plantations were established in the peninsula to supply its durable wood, esteemed for bridge building, shingles, and cabinet work, to the construction industry. Today, its timber is no longer in demand, but trees are widely planted along roadsides and in parks, where the compact, conical canopy readily distinguishes it. The leaves, which droop from the ends of twigs and are red when very young, are waxy white on their undersides. The large, fragrant, solitary flowers are produced during dry weather, and open for one day only. The fruit is a woody capsule containing up to four shiny seeds.

This species is indigenous to India (where the tree is sacred), Burma, Thailand, and Peninsular Malaysia. Various parts of the plant have entered into traditional medicine. In Indonesia, where trees are cultivated, the stamens are used to stuff pillows. The Ironwood is a member of the family *GUTTIFERAE* (see also *Garcinia mangostana*), an important component of the Malaysian rain forest with over 100 species.

Painted from a specimen in Kuala Lumpur.

Penny German 8/91

17 *Lagerstroemia speciosa* (L.) Pers.

Rose of India, Bungor

NATIVE to the seasonally dry parts of South-East Asia, India, and Australia, this spectacular tree can be seen growing wild along rivers and in open country in the lowlands throughout Malaysia, where it is also a popular ornamental. The epithet *speciosa* derives from the Latin *speciosus*, 'splendid', in reference to the tree. The showy rose-like flowers, produced in large, loose, terminal panicles nearly 0.5 metres long, display an interesting colour pattern: mauve in wild trees and pink in cultivated ones. The flowers, with their distinctive, crinkled, crêpe-like petals, soon give way to dark brown woody capsules which split when mature to reveal numerous winged seeds. The wood of the Rose of India has long been valued for its durability, colour, and texture, particularly in India where it is used for building purposes. It once served as a walnut substitute for the British timber market.

The name *Lagerstroemia* commemorates Magnus von Lagerstroem, the eighteenth-century naturalist and Director of the Swedish East Indies Company. The genus belongs to the temperate and tropical family *LYTHRACEAE*, a source of many useful dyes, ornamentals, and timber.

Painted from a specimen on the campus of the University of Malaya, Kuala Lumpur.

18 *Bauhinia purpurea* L.

Purple Bauhinia

IN contrast to the wild Bauhinias of Malaysia, which are large woody climbers of the forest fringe, the Purple Bauhinia is a small introduced tree native to India and southern China. It is distinctive in having partly split leaf blades and flowers reminiscent of an orchid, hence the vernacular names Butterfly Flower and Orchid Tree. The blooms, formed in many-flowered clusters near the branch ends, vary from pale to dark pink, to carmine, purple, and lavender. The pods are linear, flat, and up to 20 centimetres long, splitting longitudinally to release the seeds. It has been reported that seeds can be ejected 15 metres due to the explosive action of the dehiscing pod. The leaves of the plant have been used locally to poultice sores, while the flowers are said to have laxative qualities. In India, the bark is used for treating diarrhoea. Other species of this pantropical genus serve as a source of cordage, gum, dye, timber, and food.

The Purple Bauhinia is a member of the *LEGUMINOSAE* (see also *Cassia alata*). Because the paired leaf lobes or leaflets of the genus suggested a brotherly relationship, the name *Bauhinia* was chosen, commemorating Jean and Gaspard Bauhin who were sixteenth-century herbalists.

Painted from a specimen in Kuala Lumpur.

19 *Couroupita guianensis* Aublet

Cannon Ball Tree

THE Cannon Ball Tree originates from the tropical rain forests of French Guiana and the Amazon region of South America. The name *Couroupita* is a modification of a native one. Gorgeous fragrant flowers are borne on a tangled profusion of long, branch-like inflorescence stalks which arise from the trunk and lower branches of the tree. The individual flowers are made remarkable by the large waxy petals and the broad fleshy flap, fringed with stamens, which arches over the pistil. The fruits are large woody capsules, growing to 20 centimetres in diameter, resembling rusty cannon-balls and containing many seeds embedded in pulp. Uncollected fallen fruits rot and emit an unpleasant smell. The pulp is said to be eaten and made into a beverage by the natives of South America. The shell is utilized to make utensils. While the heavy fruits may inflict injury if they fall on a person, the Cannon Ball Tree is frequently cultivated in parks and gardens as a botanical curiosity.

A related species, also introduced from South America, is the Brazil Nut (*Bertholletia excelsa* Humb. & Bonpl.). Both species are members of the *LECYTHIDACEAE*, a pantropical family centred in South America where the majority of genera and species occur.

Painted from specimens in Kuala Lumpur and the Singapore Botanic Gardens.

Penny German 8/92

Garden Shrubs

WHILE true shrubs are uncommon in most rain forest habitats, their limited size makes them particularly well-suited for a variety of purposes in public and private parks and gardens. Woody shrubs are the major component of garden plant material usually seen. Once established in cultivation, they grow and continue to flower regularly, often in profusion. Because a significant portion of Malaysia's garden shrubs are of exotic origin, efforts are widening to identify and utilize more native species as garden ornamentals.

20 *Murraya paniculata* (L.) Jack

Orange Jasmine, Kemuning

THIS handsome ornamental occurs naturally in the region from India to Australia. In Malaysia, it grows wild on limestone outcrops, but is found more often in cultivation where it flowers profusely. The Orange Jasmine is an attractive shrub or small tree with shiny foliage and pleasing displays of flowers. Its large campanulate blooms, borne in small panicles, are succeeded by small red ornamental fruits which contain one or two green seeds. The flowers are often worn in the hair because of the pleasant fragrance they impart, and in Java they are used to scent cosmetics. The wood from the roots and trunk, yellow in colour, is valued by the Malays for making kris handles and walking-sticks. In Thailand, the bark of this plant was once used to make facial powder. The leaves have been used in native medicine, for example, in the treatment of venereal disease.

The Orange Jasmine is a remote relative of *Citrus* in the family *RUTACEAE* (see also *Merope angulata*). The Indian Curry Leaf, *Murraya koenigii* (L.) Sprengel, is another familiar exotic plant of Malaysian gardens from the same genus. The name *Murraya* commemorates J. A. Murray, the eighteenth-century Swedish student of Linnaeus.

Painted from a specimen on the campus of the University of Malaya, Kuala Lumpur.

Penny German 1/92

21 *Hibiscus rosa-sinensis* L.

China Rose, Bunga Raya

THE China Rose is one of the best known and most splendid ornamental shrubs in the Tropics. It has been under cultivation in eastern Asia and the Pacific Islands for so long that its origin is indeterminable. It is a shrub, 2–5 metres high, with spreading branches and leaves with toothed margins. The red corolla makes an attractive contrast with the shining dark green calyx, formed from the united sepals. In the centre of the corolla are numerous stamens, united into a tubular column around the pistil, crowned by five radiating stigmas.

Bunga Raya, Malaysia's National Flower, is cultivated in many colours and forms, including double-petalled ones. More than 1,000 different varieties have been named. Juice from the petals of this species was once used to blacken the eyebrows. It also served as shoe-blacking, hence the vernacular name Shoe Flower. The medicinal properties attributed to this plant are numerous. It has even been reported to induce abortion, and to alleviate the symptoms associated with venereal disease.

The China Rose belongs to the cosmopolitan family *MALVACEAE* (see also *Hibiscus tiliaceus*), an important source of fibre (cotton), food (okra), and ornamentals. The name *Hibiscus* is derived from the Greek and Latin term for the mallow (*Malva*) of the same family.

Painted from a specimen in the illustrator's garden, Kuala Lumpur.

22 *Bixa orellana* L.

Annatto, Kesumba

FROM the forests of tropical Central America where the Annatto grows wild, Man has carried this species to other tropical areas where it is now widely cultivated. The Annatto can grow up to 10 metres high, but it is usually maintained as a hedge or shrub. The long-stalked heart-shaped leaves and terminal clusters of pink showy flowers, combine to make it an attractive garden plant for growing singly or in borders. The coarsely hairy fruits, reminiscent of the Rambutan, split into two halves revealing many small seeds covered with a vermilion-coloured skin which yields the harmless orange-red dye called bixin (commercially known as annatto or roucou, and once an important dye for the textile industry in Europe). It is still used to colour food as it is almost tasteless. The fruit clusters themselves are esteemed by flower arrangers who use them both fresh and dried. Most parts of the plant are used in traditional medicine for treating ailments ranging from constipation and fever to asthma and leprosy.

The Annatto is the sole representative of the genus, a member of the small tropical family *BIXACEAE*. The name *Bixa* is derived from the Brazilian name for this plant, *biche*.

Painted from a specimen in the University of Malaya Botanical Gardens (Rimba Ilmu), Kuala Lumpur.

23 *Ixora javanica* (Blume) DC.

Javanese Ixora, Pecah Periuk

THE Malaysian representatives of *Ixora* comprise both native and introduced species, and are probably among the best known garden plants. The Javanese Ixora, distributed throughout the Indo-Malesian region, grows as a shrub under cultivation, but attains heights of over 6 metres when growing wild in the forest. The flowers, each with a long slender corolla tube and four widely spreading petals, occur in crowded flattened clusters, 5–10 centimetres across, at the ends of the branches. The flowers exhibit considerable variation in colour: yellow, red and yellow, pink or orange darkening to red. Garden varieties flower throughout the year, while the forest species are seasonal. They appear to be pollinated by butterflies.

The related species, *Ixora chinensis* Lam. (Chinese Ixora) and *I. coccinea* L. (Indian Ixora), both exotic garden shrubs in Malaysia, are medicinal. In addition to the natural species of *Ixora*, there are numerous horticultural hybrids and varieties, many of unknown origin. This pantropical genus, comprising about 400 species, is a member of the family *RUBIACEAE* (see also *Gardenia carinata*). The name *Ixora* is a Portuguese rendering of the Sanskrit name of the god Siva.

Painted from a specimen in Kuala Lumpur.

Penny German 2/93

Herbaceous Plants

HERBACEOUS plants are typically non-woody and of low height. They form an important component of the ground vegetation in tropical forests. In the interior of the lowland forest, they sometimes appear to be absent, but become more luxuriant in clearings and along streams. The dicot and monocot plants, which constitute a major portion of Malaysia's herbaceous flora, show great variety in the form of their underground parts, foliage, and flowers. They show less diversity in number of species, although this varies from place to place.

24 *Tacca integrifolia* Ker-Gawler

Black Lily, Keladi Murai

THE Black Lily is a relatively common tuberous herb which grows in the deep shade of Malaysia's lowland and hill forests. The long-stalked tapered leaves and erect flowering stalks are produced from a subterranean stem, and can grow up to 1 metre in height. The dull dark purple flowers, arranged in clusters of five or more, are surrounded by large purplish-black to white leaf-like bracts, with a number of pendant thread-like bracts. The entire inflorescence is reminiscent of a bat, giving rise to the vernacular name Bat Flower. Each fruit, a pendulous purplish-black keeled berry, contains numerous small brown ribbed seeds. The Malays of Peninsular Malaysia have reportedly used the pulped tubers of the Black Lily as a poultice for certain skin irritations.

A related plant of sandy coasts, *Tacca leontopetaloides* (L.) Kuntze, the Tahiti Arrowroot or Lukeh, is the source of an edible, bitter starch. Both species are members of the *TACCACEAE*, a small family of tropical monocotyledonous herbs. The name *Tacca* is believed to derive from a native Ambon one.

Painted from a specimen in secondary forest at the Forest Research Institute Malaysia, Kepong, Selangor.

25 *Amorphophallus campanulatus* Blume

Elephant Foot Yam, Loki

THE Elephant Foot Yam is a herbaceous perennial with a large underground corm from which arises a single, much-divided leaf, growing to over 1 metre in height. Once the leaf has withered away, the inflorescence emerges bearing numerous tiny yellow male and female flowers embedded in a central spike, the spadix, which terminates in a deeply wrinkled conical appendage. From the base of the spike develops a large, leaf-like, reddish-brown spathe. The newly opened inflorescence smells of rotting meat, attracting carrion insects which effectively pollinate the flowers. The corms of this plant are eaten for their starch, but only after they are washed and boiled to remove chemical irritants.

This species is a common herbaceous plant of open places and secondary growth in the lowlands throughout South-East Asia. Although not reaching the dimensions of its colossal Sumatran relative, *Amorphophallus titanum* Beccari (its corm alone can weigh up to 50 kilograms!), Loki is often cultivated as a horticultural curiosity.

The name *Amorphophallus* derives from the Greek *amorphos*, 'misshapen', and *phallos*, 'penis', an allusion to the inflorescence. The genus is a member of the large, predominantly tropical and subtropical family *ARACEAE* (see also *Arisaema filiforme*), a valuable source of food plants in the Tropics.

Painted from a specimen at the Department of Botany, University of Malaya, Kuala Lumpur.

26 *Trimezia martinicensis* Herbert

Forenoon Yellow Flag

ORIGINATING from tropical America, the Forenoon Yellow Flag has successfully established itself in the Malaysian flora where it grows as a weed in open grassy places. The bulbous underground stem gives rise to short, narrow, grass-like leaves, and an inflorescence stalk which bears a succession of flowers which open in forenoon and wither by midday. The sepals and petals of the flower, technically called perianth segments, are similar in appearance (yellow with a brownish stain at the base) and occur in two series of three segments each, the outer series being larger than the inner series. The name *Trimezia* is, in fact, derived from the Greek *treis*, 'three', and *megas*, 'great', alluding to the outer perianth segments. The fruit is a pale green capsule which drys and splits, releasing the reddish-brown seeds.

This species belongs to the cosmopolitan family *IRIDACEAE*, none of which are native to Malaysia, although several exotic genera (*Iris, Gladiolus*, and *Crocus*) are cultivated as colourful ornamentals. In some regions, the Forenoon Yellow Flag is valued as an ornamental and is cultivated in greenhouses or planted outdoors. Certain other South American species of *Trimezia* are known for their laxative properties.

Painted from a specimen along the old Genting Highlands Road, Selangor.

27 *Orthosiphon aristatus* (Blume) Miq.

Cat's Whiskers, Misai Kucing

THIS slender herb of South-East Asia is both indigenous to and cultivated in Malaysia. The four-angled stems, growing to 30–60 centimetres high, bear narrow toothed leaves arranged in pairs. Both the leaves and stems are pleasantly aromatic due to the presence of glands containing volatile oils within their tissues. The flowers, arranged in whorls in lax terminal racemes, consist of a slender corolla tube from which protrude four stamens and one stamen-like style. Both white- and purple-flowered forms of this species exist, making it a popular bedding plant for tropical gardens. The leaves of the Cat's Whiskers have long been utilized in Malaysia, Indonesia, and the Philippines to treat a wide range of urinary ailments, as well as arteriosclerosis, gout, and rheumatism. In research conducted in Europe, the plant was determined to be a superior diuretic.

This species, the sole representative of the genus in Malaysia, is a member of the *LABIATAE*, a large cosmopolitan family long valued for its medicinal and ornamental uses, and many still of great importance as a source of flavourings and scents. The name *Orthosiphon* is derived from the Greek *orthos*, 'straight', and *siphon*, 'pipe', in reference to the corolla.

Painted from a specimen in the University of Malaya Botanical Gardens (Rimba Ilmu), Kuala Lumpur.

Sea-shore Plants

SEA-SHORE plants grow in the narrow strip of vegetation which develops immediately behind the beach. This strand vegetation is never found at any great distance from the sea, unless planted inland, and the plants found here, while botanically diverse, are all adapted for survival in this adverse environment. Because the species occupying this habitat have seeds and fruits that float in sea water, and are dispersed by it, many of them are widely distributed on tropical sea-shores throughout the Indo-Malesian and Pacific regions.

28 *Hibiscus tiliaceus* L.

Sea Hibiscus, Bebaru

THE Sea Hibiscus is a small to medium sized tree inhabiting sandy and rocky sea-shores throughout Malaysia and the Tropics. It is also frequent along the fringes of mangrove forest, where the spreading, much-branched crown can form an almost impenetrable thicket. The tree is easily recognized by the green heart-shaped leaves and the attractive flowers which develop in clusters near the branch ends. The flowers comprise overlapping yellow petals with a maroon spot at the base of each. In the centre of the flower is a tube of fused stamens crowned by crimson stigmatic lobes. The flowers last only one day, fading to pinkish-yellow after falling.

The Sea Hibiscus grows well inland, far from its natural habitat, where it can be seen planted along roads and in parks. The fibre from the inner bark is made into cordage, nets, coarse fabrics, and tow for caulking ships. The soft wood has been used for planking, and for constructing boats and household implements. All plant parts, including the flowers, are of local medicinal importance.

The Sea Hibiscus is classified under the family *MALVACEAE* (see also *Hibiscus rosa-sinensis*). The epithet *tiliaceus* refers to the shape of the leaves, which resembles that of the temperate tree, *Tilia*.

Painted from a specimen on the coast near Kuala Sedili Kecil, Johor.

29 *Scaevola sericea* Vahl

Sea Lettuce, Ambong Ambong

THE Sea Lettuce is a familiar shrub of sandy and rocky sea-shores throughout Malaysia. This bushy plant grows to 6 metres in height and produces large, succulent, waxy leaves with recurved margins. The white flowers, often tinged pale lilac, occur in clusters which develop from the leaf axils, and are produced all year round. The hairy five-lobed corolla tube is unique in being split to the base along one side, exposing the slender style and stamens. The fruit is a fleshy berry-like drupe, ripening white, and containing one or two seeds. The seeds can retain their viability for long periods in sea water but germinate only in freshwater, such as rain water. The juice from the fruits is said to soothe sore eyes. The fleshy leaves are made into a poultice for headache and swollen limbs, and are also reported to be eaten as a cure for indigestion. The durable wood was once used for the treenails in boat making.

The Sea Lettuce is widespread throughout the Indo-Malesian region and the Pacific Islands. It is the sole Malaysian representative of the *GOODENIACEAE*, a small plant family of over 400 species of mainly Australian origin.

Painted from a specimen on Pulau Pangkor, Perak.

30 *Pandanus odoratissimus* L. f.

Sea-shore Screw Pine, Pandan Laut

THE phrase, 'strange palm-like tree with pineapple-like fruits', aptly describes the common Sea-shore Screw Pine found on almost all tropical shores from the Indian Ocean eastward to the Pacific Islands. It is widespread on Malaysian coasts, and is even cultivated inland. The much-branched trunk of the plant is supported at the base by heavy stilt roots. The stiff sword-like leaves grow to 1.5 metres long, and are armed with sharp prickles. The male and female flowers are produced in dense spikes on separate plants. The fragrant male flowers (hence the epithet *odoratissimus* from the Latin meaning 'very fragrant') develop in branched, pendulous spikes. The female flowers are tightly aggregated in a globular 'head' which develops into the 'fruit', comprising fused woody drupes. The leaves provide material for matting, baskets, cordage, and hats, and all parts of the plant, including the aerial roots, have been used in native medicine. In Polynesia, the edible kernels are a principal food source.

The genus *Pandanus* of the family *PANDANACEAE* is well-represented in Malaysia by both introduced and wild species, the latter occupying habitats from sea-level to the high mountains. The popular name, Screw Pine, alludes to the spiral arrangement of its leaves.

Painted from a specimen near Port Dickson, Negeri Sembilan.

31 *Vitex ovata* Thunb.

Creeping Vitex, Lenggundi

A creeping plant of sandy sea-shores, the Creeping Vitex is distributed throughout tropical Asia, Australia, and Polynesia. In Peninsular Malaysia, it is commonly seen along the east coast, sending out long woody runners across the sand. From the rooted stems arise short, upright, leafy shoots (up to 40 centimetres high) with grey-green leaves, which are rather pungent when crushed. The specific epithet *ovata* derives from the Latin *ovatus*, 'egg-shaped', in reference to the leaves. The flowers are produced in crowded inflorescences at the ends of the upright shoots. The two-lipped deep blue to violet corolla gives way to a small round berry surrounded by an enlarged calyx, which eventually splits along one side.

Some botanists regard this species as a variety of *Vitex trifolia* L., a well-known village shrub of some importance in native medicine: its leaves are said to have been burned by the Malays to drive away mosquitoes—and even evil spirits! Lenggundi is one of several Malaysian species of *Vitex*, the majority of which are evergreen trees, belonging to the family *VERBENACEAE*. The name *Vitex* derives from the ancient Latin name of the Chaste Tree (a species of *Vitex*) or some similar shrub.

Painted from a specimen on the beach at Dungun, Terengganu.

32 *Cerbera manghas* L.

Red-eyed Cerbera, Pong Pong

THE Red-eyed Cerbera is a characteristic plant of all rocky and sandy sea-shores throughout the region from Thailand and Peninsular Malaysia eastward to Papua New Guinea. This small tree produces a dense crown with glossy spatulate leaves arranged in clusters at the ends of the branches. The stems exude a sticky white latex when cut. The fragrant flowers, borne in lax terminal panicles, have white funnel-shaped corollas with red centres. The large, hard, drupaceous fruit, often twinned with another, is glossy green and red when ripe and highly reminiscent of a mango fruit, hence the epithet *manghas*, the Portuguese word for the mango. Many parts of the plant are medicinal; the seeds are reported to contain the heart poison cerberin. Consumption of the poisonous fruit (with seeds) is said to have been a method of suicide in the Marquesas Islands in the South Pacific.

The name *Cerbera*, alluding to the poisonous attributes of the plant, derives from Cerebus, the name of the three-headed dog which, according to Greek mythology, guarded the entrance to the nether world. The Red-eyed Cerbera is a member of the largely tropical family *APOCYNACEAE*, its species constituting a source of many drugs and poisons, ornamentals, and timber.

Painted from a specimen on Pulau Perhentian off the Terengganu coast.

Gymnosperms

UNLIKE the angiosperms which produce their seeds within ovaries that eventually develop into fruits, the seeds of gymnosperms lie exposed on the scales of cones or other structures. The gymnosperms predated the flowering plants in evolutionary history, and include some very primitive seed plants, representatives of which still exist today. The gymnosperms account for less than 1 per cent of Malaysia's indigenous flora. Their habitats range from sandy beaches and limestone formations to lowland forests and high mountains.

33 *Gnetum gnemon* L.

Meninjau

MENINJAU is indigenous to the Indo-Malesian region and the Pacific Islands as far east as Fiji. Several varieties of this species exist in Malaysia, the most familiar being the tree form (variety *gnemon* L.) which is occasionally cultivated in villages and parks, and grows wild on some rocky headlands and islands. The other varieties, all shrubs, are found in lowland and mountain forest throughout the country.

The tree variety has a characteristically narrow pyramidal crown and nodular trunk with drooping branches and broad leaves. The erect spike-like male and female cones, with their tiny 'flowers' arranged in whorls, are produced on separate plants. The plump seeds have skins which ripen from yellow to orange-red and contain a starchy edible kernel which is processed into food. A useful fibre is extracted from the inner bark. The roots are said to be an antidote for poison and a remedy for malaria.

In addition to Meninjau, several other species of *Gnetum* are represented in the Malaysian flora, all forest climbers. They belong to the *GNETACEAE*, a remarkable tropical family of gymnosperms with angiospermous traits, notably the broad reticulate-veined leaves and 'stamens' in the male cones. The name *Gnetum* and the epithet *gnemon* derive from the Moluccan name for the tree.

Painted from a specimen on the campus of the University of Malaya, Kuala Lumpur.

34 *Cycas rumphii* Miq.

Sea-shore Cycad, Paku Laut

THE cycads derive their name from *kykas*, the Greek word for an African plant with a palm-like growth habit. Like most cycads, the Sea-shore Cycad can easily be mistaken for a palm tree or even a tree fern. It is a robust plant with a columnar trunk, 2–5 metres high, made rough by the persistent leaf bases. The trunk is crowned by spreading, leathery, pinnate leaves (with 50–60 leaflets each) that can attain lengths of up to 2 metres. Individual trees are unisexual. In male plants, overlapping pollen-bearing scales are united into an impressive golden-brown cone 30 centimetres high. In female plants, the cones (shown here) comprise a loose aggregation of pendulous seed-bearing leaves (sporophylls) arising from the end of the stem. The starchy (and poisonous) seeds and young leaves are edible when cooked, while *sago* can be extracted from the stem. Further use is made of the plant in traditional medicine; the pollen is said to be narcotic.

The Sea-shore Cycad grows naturally along the coast of Malaysia and elsewhere throughout the Indo-Malesian region. It belongs to the very primitive family of gymnosperms, the *CYCADACEAE*, whose ancestors flourished with the dinosaurs more than 200 million years ago.

Painted from a specimen in strand vegetation along the Terengganu coast.

35 *Podocarpus imbricatus* Blume

Malaysian Yellow Wood, Cucur Atap

THE Malaysian Yellow Wood is a tropical conifer indigenous to the montane forests of Indo-China and the Malesian region. It is frequent in Malaysian forest at 700–1 800 metres elevation, but is occasionally planted in the lowlands. Trees can reach heights of up to 30 metres in the forest. Their crowns are conical and rather straggly. Two types of leaves are formed: linear ones arranged in feather-like sprays at the ends of shoots, and short needle-shaped ones arranged spirally on branches. The male cones, each measuring 1 centimetre long and comprising a series of tiny scales bearing pollen sacs, occur in small clusters on the branches. The female cones (shown here) develop on separate trees and consist of a single globose seed seated atop a swollen fleshy receptacle.

The name *Podocarpus* is derived from the Greek *podos*, 'foot', and *karpos*, 'fruit', in reference to the inflated seed stalk. The pinkish or yellowish wood of this tree is even-grained and valued for furniture and cabinet work. The Malaysian Yellow Wood is one of ten species of the family *PODOCARPACEAE* in Malaysia. The family comprises chiefly natives of the Southern Hemisphere.

Painted from a specimen in the arboretum of the Forest Research Institute Malaysia, Kepong, Selangor.

Montane Plants

As one ascends the forested slopes of the mountains and the last of the hill dipterocarp vegetation is left behind, the montane plants gradually make their appearance. Nearly constant moist conditions here favour the growth of lush vegetation, from the lowly mosses and lichens to the luxuriant growth of herbaceous plants and ferns, vines, shrubs, and trees. Because of cooler temperatures prevailing at these altitudes, Malaysian mountain forests show affinities with temperate floras which have made intrusions from both the north and the south.

36 *Impatiens oncidioides* Ridley

Golden Balsam

Considered one of the most beautiful of the local balsams, the Golden Balsam is endemic to Peninsular Malaysia where it grows wild in damp protected places on the Main Range, often adjacent to streams and frequently in muddy ground. It is a succulent herb, growing not more than 50 centimetres tall, with a cluster of lax thin leaves at the top of the fleshy stem. One or two yellow flowers, turned 180 degrees on their stems, are produced on an inflorescence stalk. As in all balsam plants, one sepal has been modified into a backward-projecting spur which produces nectar. The fleshy capsular fruit is also unusual in forcibly discharging the seeds through an explosive, twisting action when ripe. The Latin word for 'impatient' alludes to this character and gives the genus its name.

Several other species of *Impatiens* occur in Malaysia, a number of them endemic to the peninsula, and some found exclusively on limestone. The Garden Balsam (*I. balsamina* L.) is a popular bedding plant in the lowlands. The Golden Balsam is classified under the *BALSAMINACEAE*, a family of over 800 species of largely tropical herbs, few of which are actually cultivated.

Painted from a specimen at the base of Gunung Brinchang, Cameron Highlands, Pahang.

37 *Begonia pavonina* Ridley

Peacock Begonia

THE Peacock Begonia is without a doubt the most striking of Malaysia's montane begonia species, with its satin-like sheen and iridescent leaves of green-peacock blue (hence the epithet *pavonina* from the Latin, *pavo*, 'peacock') tinged red beneath. These colours, intensifying in plants growing under deep shade, are regarded as an adaptation designed to increase photosynthesis under low light conditions. A long slender inflorescence stalk grows from the base of the plant and bears several unisexual flowers: male flowers above and female flowers below. The fruit, a three-winged capsule, splits open to expose numerous minute seeds. The Peacock Begonia grows in mountain forest, at 1 200 metres elevation, on the Main Range of Peninsular Malaysia, under shade and usually in wet ground. Many of Malaysia's native begonias are montane plants, and several species are found growing only on limestone.

All begonias belong to the family *BEGONIACEAE*, with over 900 species world-wide, and the source of important garden plants appreciated for both their flowers and foliage. The genus *Begonia* has few uses other than as a rich source of ornamentals, and over 10,000 hybrids and cultivars are known today. The name commemorates Michel Begon, patron of botany and former Governor of French Canada.

Painted from a specimen at the base of Gunung Beremban, Cameron Highlands, Pahang.

38 *Arisaema filiforme* Blume

THIS herbaceous perennial aroid of the family *ARACEAE* (see also *Amorphophallus campanulatus*) produces a subterranean tuber from which arises two leaves. The leaves consist of three pointed leaflets at the end of a fleshy reddish petiole that grows to 50 centimetres in height. The attractive inflorescence is borne on an erect stalk of similar length. The coloured leaf-like spathe envelopes the spike-like spadix and bends forward, forming a hood over it. The minute unisexual flowers are closely packed on the spadix which terminates in a long slender appendage (hence the epithet *filiforme* from the Latin *filum*, 'thread', and *forma*, 'form') protruding above the mouth of the spathe tube. From the female flowers develop bright orange-red berries on a short exposed spike.

This plant is native to Peninsular Malaysia, where it inhabits dense forest, often near streams, in the mountains. No economic uses for it are recorded, but the tubers of an unidentified *Arisaema* from a similar habitat in the peninsula are reported to have been used in a wash for malaria victims. The name *Arisaema*, alluding to the coloration found in certain kinds, is derived from the Greek *arisaron*, a kind of 'arum', and *haima*, 'blood'.

Painted from a specimen at the base of Gunung Brinchang, Cameron Highlands, Pahang.

Orchids

CHARACTERISTIC of, but not limited to, the Tropics, the orchids (Bunga Orkid) are the largest group of angiosperms in Malaysia and can be found in all habitats, from sea-level to the mountains. Man's fascination with these highly specialized plants can be attributed to the near limitless variation seen in their vegetative and floral structures. Orchids belong to the largest family of flowering plants, the *ORCHIDACEAE*, with estimates of up to 30,000 species world-wide, not including the myriad garden hybrids.

39 *Bulbophyllum skeatianum* Ridley

Skeat's Bulbophyllum

ENDEMIC to Peninsular Malaysia, Skeat's Bulbophyllum is not a common species; it is known from only four or five locations in the mountains between 1 200 and 1 800 metres elevation. A long creeping rhizome, attached to the moss-covered limbs of trees, bears a series of angled oval pseudobulbs, each terminated by a single fleshy leaf. The name *Bulbophyllum*, alluding to the pseudobulbs and the leaves that sprout from them, derives from the Greek *bolbos*, 'bulb', and *phyllon*, 'leaf'. From the base of the pseudobulb arises a slender purple scape bearing a group of about ten flowers arranged like a fan. The two petal-like lateral sepals are unique in being twisted at the base and fused together along their upper edges down to their tips. The upper sepal and petals are much smaller and coloured red-purple.

Over 100 species of *Bulbophyllum* are native to Malaysia, making it the largest genus of orchids in the country. Despite their graceful and attractive flowers, the majority of species are rarely cultivated. With over 1,000 species distributed in the New and Old World Tropics, Japan, and New Zealand, the genus ranks as one of the largest of the orchid family.

Painted from a specimen on trees along the forested slopes of Gunung Beremban, Cameron Highlands, Pahang.

Penny Cameron 12/92

40 *Phalaenopsis cornu-cervi* (Breda) Bl. & Rchb. f.

LIKE all members of the genus *Phalaenopsis*, this species is an epiphytic orchid. It has a short stem, lacks pseudobulbs, and produces narrow, leathery, pale green leaves. The flowers arise in two ranks from a branched inflorescence stalk which is flattened and bears conspicuous alternate bracts. Each star-shaped flower is coloured light yellow-green with red-brown blotches, and reveals a lip of rather complex structure.

Phalaenopsis cornu-cervi, a native of Thailand, Peninsular Malaysia, Sumatra, and Borneo, is found on forest trees at low elevations. It is known to hybridize naturally with a related species, the mauve-coloured *P. violacea* Teijsm. and Binnend., in places where they grow together. The genus *Phalaenopsis* comprises 70 species distributed from the Indo-Malesian region to Australia and Taiwan. Nearly all produce attractive flowers, in both colour and form, and several species are popular garden orchids. The name *Phalaenopsis* comes from the Greek *phalaina*, 'moth', and *-opsis*, 'resembling', in reference to the form of the flower. The vernacular name, Moth Orchid, is applied to certain species of the genus.

Painted from a specimen at the Orchid Garden, Kuala Lumpur.

41 *Paphiopedilum barbatum* (Lindley) Pfitzer

Lady's Slipper, Bunga Kasut

THE name *Paphiopedilum* is derived from *paphia*, which relates to Paphos, the chief centre of worship of the goddess Venus, and *pedilon*, 'slipper', alluding to the form of the lip. The Sandal of Venus, as this plant is sometimes called, is a terrestrial orchid with conspicuously mottled leaves, in two shades of green, arising from a subterranean stem. The slender scape grows to about 25 centimetres in height and bears one (rarely two) flower 8 centimetres across. The perianth parts—a broad upper sepal, two lateral petals, the pouched lower petal or lip, and two united lateral sepals—show considerable variation in coloration. The two lateral petals are fringed with hair (hence the epithet *barbatum* from the Latin *barba*, 'beard') and have little warts along their margins.

This species, first discovered in 1838 on Gunung Ledang in Johor, has been extensively hybridized and cultivated in Europe. It grows naturally in cool shady areas in the hills, from Peninsular Malaysia northward into Thailand. Zealous plant hunters continue to strip the forests of these and other Lady's Slipper orchids as they are of great horticultural value in northern countries. For a long time, the orchids of this genus were known under its previous name, *Cypripedium*.

Painted from a specimen in the University of Malaya Botanical Gardens (Rimba Ilmu), Kuala Lumpur.

42 *Spathoglottis aurea* Lindley

THIS beautiful terrestrial orchid is frequently seen along roadsides and in open grassy areas in hills and mountains throughout Peninsular Malaysia, Sumatra, and Java. The long plaited leaves and large golden yellow flowers of this species are unmistakable. The narrow, lanceolate, purple-tinged leaves arise from pseudobulbs, and a tall slender scape (up to 60 centimetres high) terminates in a raceme of flowers which open in succession over a long period of time. The sepals and petals, spreading widely in the opened flower, are similar in colour and shape, with the exception of the three-lobed lip which is streaked or spotted crimson.

This highland species, originally discovered on Gunung Ledang, Johor, is difficult to cultivate at low elevations. In contrast, the more widespread species, *Spathoglottis plicata* Bl. (also illustrated), often found growing with *S. aurea* in open places in the hills, has long been cultivated in the lowlands, and numerous hybrids and varieties of it exist. Its flowers are typically coloured mauve, but bright purple- and white-flowered varieties are also known. The name *Spathoglottis* derives from the Greek *spathe*, 'spathe', and *glotta*, 'tongue', referring to the form of the lip of the flower.

Painted from a specimen along the old Genting Highlands Road, Selangor.

Fruit Trees

THE tropical fruit trees of Malaysia are represented by indigenous species and introductions, early and modern, from other tropical countries, both from within and outside South-East Asia. Many of the commonly cultivated tropical fruit trees of home gardens and orchards have originated from wild types, relatives of which can still be encountered in the forest. Through their diverse kinds, and their many uses, fruits have long played a significant role in the cultures and traditions of Malaysia and the surrounding region.

43 *Eugenia malaccensis* L.

Malay Apple, Jambu Bol

THE genus *Eugenia* is Malaysia's largest genus of flowering plants with nearly 200 wild species, mostly trees, occurring in the peninsula alone. The Malay Apple is one of a few species of introduced plants of the genus—its exact origin is uncertain—cultivated in villages and gardens. It is also the most easily recognized. The tree, which can grow to 18 metres high, has a cylindrical crown and large leathery leaves which droop from the branches. The large, bright, crimson-pink flowers are borne in compact clusters which arise from the leafless stems. They resemble powder puffs because of the densely packed, slender stamens. The flowers are succeeded by large oblong to pear-shaped berries, ripening dark red, purplish-yellow, or yellow-white. The flesh of the berries is white, juicy, and rather insipid, and can be eaten raw or cooked, and made into preserves.

Other familiar Eugenias are the edible Jambu Air (*E. aquea* Burm. f.) and the Clove Tree (Cengkeh, *E. aromatica* Kuntze). The name *Eugenia* commemorates Francois Eugène, Prince of Savoy, the eighteenth-century patron of science and literature. The genus belongs to the tropical and warm temperate family *MYRTACEAE*, an important source of timber, food plants, spices and medicinal oils, and handsome ornamentals.

Painted from a specimen on the campus of the University of Malaya, Kuala Lumpur.

44 *Nephelium lappaceum* L.

Rambutan

The Rambutan is an untidy broad-crowned tree growing up to 30 metres in height, with dark green leaves composed of 2–4 pairs of leaflets. The two kinds of inconspicuous flowers, male and bisexual, are produced on separate trees in branched inflorescences. The bisexually flowered inflorescences develop up to several dozen fruits. The skin of the fruit, ripening red or yellow, is covered with soft spines, and encloses a white, translucent, grape-like flesh containing a single fertile seed. The name *Nephelium* is, in fact, derived from the Greek *nephele*, 'cloud', alluding to the flesh of the fruit. The quality of the flesh varies considerably, with both sweet and sour varieties being known. The oily seeds are sometimes eaten roasted, or are made into rambutan tallow. Many parts of the plant have entered into traditional medicine.

While many cultivars of the Rambutan have been developed and introduced into tropical countries, the plant's natural distribution ranges from southern China southward to Indo-China, Malaysia, Indonesia, and the Philippines. The Rambutan and its allies, including the Litchi, Longan, and Pulasan, are classified in the largely tropical family, *SAPINDACEAE*, a source of edible fruits, timber, oil-seeds, and ornamentals.

Painted from a specimen in the illustrator's garden, Kuala Lumpur.

45 *Manilkara zapota* (L.) P. Royen

Sapodilla, Ciku

THE Sapodilla originates from the Central American Tropics and is now cultivated throughout the tropical world. It first arrived in South-East Asia with the Spaniards who carried the fruits to the Philippine Islands. This evergreen tree, 10–12 metres high, has a rather shabby appearance. The leaves are crowded at the ends of the branches, and from their axils arise small white flowers. The fruit is a round or oblong berry with a scurfy skin and juicy sweet flesh containing a number of flattened black seeds. The Sapodilla produces flowers and fruits all year long, and is commonly planted as a fruit tree in orchards and village gardens throughout the country.

In Central America, the trees are tapped for the white latex known as chicle, once used in the manufacture of chewing gum and giving rise to the vernacular name Chewing Gum Tree. The ancient Aztecs used the dried latex for the same purpose. The bark and green fruits are high in tannin, and have been employed in village medicine to treat diarrhoea.

The Sapodilla is a member of the *SAPOTACEAE*, a family of mainly tropical trees and shrubs, and an important source of edible fruits, oils, timber, and gutta-percha. The name *Manilkara* derives from a Malabar vernacular one.

Painted from a specimen in a home garden in Petaling Jaya, Selangor.

46 *Artocarpus integer* (Thunb.) Merr.

Cempedak

THE Cempedak is widely distributed across South-East Asia, from Burma eastward to Irian Jaya, where it occurs as both a wild and cultivated tree. It is an evergreen, growing up to 20 metres tall, with leathery leaves and round or oblong male and female flower heads comprised of densely packed unisexual flowers. The cylindrical, warty, compound fruits—growing to 35 centimetres in length—are derived from the swollen female flower heads which arise from the main branches and trunk. The male heads are formed on shoots in the periphery of the canopy, and fall off after pollination. The mature fruits, which contain many large seeds each enclosed by a soft mushy jacket, produce a strong, highly pervasive smell. Wild trees of the variety *silvestris* Corner, or Bangkong, are said to be odourless. The flesh is sweet and can be eaten fresh or cooked, as can the young fruits and leaves, and seeds.

The name *Artocarpus* is derived from the Greek *artos*, 'bread', and *karpos*, 'fruit', and refers to the bread-like quality of the fruit of *Artocarpus altilis* (Parkinson) Fosb., the Breadfruit, when baked. The genus is classified under the predominantly tropical family *MORACEAE*, a source of many important fruit trees and ornamentals, timber, and rubber.

Painted from a specimen in a home garden in Kuala Lumpur.

47 *Garcinia mangostana* L.

Mangosteen, Manggis

THE Mangosteen is known only from cultivation, although there are reports of wild trees growing in Peninsular Malaysia. From South-East Asia, the crop has spread to other tropical areas. Mangosteens are regarded as the supreme fruit of the Tropics, and can be found in markets as far away as Europe and the United States. The Mangosteen is a small tree with a dense pyramidal crown of dark heavy foliage which nearly reaches the ground. All parts of the plant exude a yellow latex when damaged. Female flowers with fleshy broad petals and sterile stamens (staminodes) are produced at the ends of the branches. Male flowers are unknown. The globose purplish-brown fruits, crowned by the persistent stigmatic lobes, are able to develop without fertilization. Under the dark red fibrous flesh of the fruit are up to eight succulent, pure white segments of the most delectable flavour. The rind is medicinal, and is also used to tan leather and to dye black.

The genus *Garcinia*, named in honour of Laurent Garcin, the French physician and naturalist of the eighteenth century, is represented in Malaysia by nearly 50 species, nearly all wild forest trees and several of minor economic importance. It is classified under the *GUTTIFERAE* (see also *Mesua ferrea*), a large and varied family of the Old World Tropics.

Painted from a specimen in a home garden in Kuala Lumpur.

Gingers

THE gingers are a large and diverse group of mostly herbaceous tropical monocots characterized by their usually aromatic parts, especially the subterranean rhizomes. Included here are the common ginger and a number of other plants valued for their use as spices, scents, and medicines. The attractive foliage and orchid-like flowers of many ginger plants make them popular as cultivated garden ornamentals. All gingers belong to the predominantly Asiatic family *ZINGIBERACEAE*, comprising about 1,200 species.

48 *Alpinia galanga* (L.) Sw.

Greater Galangal, Lengkuas

THE Greater Galangal, a native of the Indo-Malesian region, is widely cultivated throughout South-East Asia for its aromatic rhizome, which is used as a spice and as an ingredient in native medicine. In Malaysia, it is frequently planted around villages from where it often escapes and spreads to disturbed ground and waste places. Several stout leafy stems, growing up to 2 metres tall, arise from the thickened rhizome. The inflorescences are borne at the ends of these stems and comprise numerous short branches with 3–5 flowers each. Each flower consists of a short tubular calyx, pale green petals (the uppermost one hooded), and a notched white lip (the petal-like staminode) streaked with pink or red. A single massive anther protrudes from the throat of the corolla tube. The fruit is a round red capsule containing a few seeds. The flowers of the Greater Galangal are eaten fresh, in salads, by the Javanese.

The genus *Alpinia* comprises over 230 species distributed throughout the warm parts of Asia and the Pacific. In the Tropics, they are popular garden plants because of the handsome foliage and attractive flowers they produce. The name *Alpinia* commemorates Prosper Alpinus, the seventeenth-century Italian botanist.

Painted from a specimen in the illustrator's garden, Kuala Lumpur.

49 *Etlingera elatior* (Jack) R. Sm.

Torch Ginger, Kantan

THE Torch Ginger grows wild in lowland forest throughout Malaysia and Sumatra, where its tall torch-like inflorescence distinguishes it readily from other herbaceous plants. It is frequently cultivated in gardens for the young flower shoots and half-ripe fruiting shoots which are added to curries, and for the ripe seeds which are eaten raw. Further use is made of the stems for matting and paper-making, and of the leaves and fruits in local medicinal remedies for alleviating earache and for cleansing wounds.

The Torch Ginger is a perennial herb with arching leafy stems, up to 5 metres tall, arising from an underground rhizome. Each stem produces about 18 pairs of leaves, the largest of which can exceed 80 centimetres in length. The pink flowers, with only the yellow or white margined lips visible, develop from the axils of bracts which are tightly arranged in a dense, cone-shaped head at the end of a leafless stalk about 1 metre high. A series of larger bracts, overlapping and petal-like, spread out from the base of the head. The decorative and brilliantly coloured inflorescences are often used in floral arrangements. This species was previously classified under the genera *Phaeomeria* and *Nicolaia.*

Painted from a specimen in Kenny Hills, Kuala Lumpur.

50 *Zingiber spectabile* Griffith

Black Gingerwort, Tepus Tanah

The Black Gingerwort is a perennial herb, not uncommon in the moist lowland forests of Peninsular Malaysia where it is native. The underground rhizome produces tall thin-leaved stems up to 2 metres high, and leafless fertile stems up to 50 centimetres high, terminated by the inflorescence. The cylindrical cone-shaped inflorescence, the largest and most conspicuous of any Malaysian species of *Zingiber* (hence the epithet *spectabile* from the Latin *spectabilis*, 'showy'), is composed of broad overlapping bracts which are at first yellow and then turn red. A single short-lived flower with pale yellow petals and a purple lip, arises from the axil of each bract. Only a few flowers are produced by the inflorescence at any one time. The plant has been used medicinally—the pounded leaves are applied as a poultice to inflamed eyes—and to flavour food.

The Black Gingerwort is related to the ginger of commerce, *Zingiber officinale* Roscoe (Halia), a valuable spice plant of indeterminate origin, known from tropical Asia since antiquity, and now cultivated throughout the tropical world. The name *Zingiber* is a corruption of the Sanskrit word meaning 'horn-shaped', in reference to the rhizome.

Painted from a specimen in hill forest near Genting Sempah, Selangor.

51 *Kaempferia pulchra* Ridley

Limestone Kaempferia

THIS attractive and floriferous ginger is known only from the Langkawi Islands of northern Peninsular Malaysia and southern Thailand where it grows between the cracks of limestone rock. Under cultivation, it flourishes in sandy soil under shade, and is particularly well-suited for rock gardens. The dark green and variegated leaves of the Limestone Kaempferia develop from an underground aromatic rhizome, and are spread out flat. A short inflorescence stalk, its base enclosed by the leaf sheaths, gives rise to flattened flowers nearly 5 centimetres across. The two large staminodes and upper bilobed lip give the appearance of a four-petalled, lilac-coloured flower. The actual petals, rather small and inconspicuous, are rolled backward behind the flower. The tip of the anther protrudes slightly beyond the mouth of the white corolla tube.

Two other species of *Kaempferia*, widely cultivated in Malaysia and elsewhere in South-East Asia, are Cekur (*K. galanga* L.) and Kencur (*K. rotunda* L.), both valued as food flavourings, and for their medicinal properties. The name *Kaempferia* commemorates the seventeenth-century German physician and botanist, Engelbert Kaempfer.

Painted from a specimen in a home garden in Kuala Lumpur.

Penny Berman 11/92

Pitcher Plants

THESE carnivorous plants of the small Old World tropical family *NEPENTHACEAE*, comprising a single genus, *Nepenthes*, are a marvel of the plant kingdom. The pitchers, variously coloured and equipped with nectaries, develop at the ends of the leaf tendrils, and are receptacles for the secretion of enzymes which dissolve insects that have fallen into the digestive fluid. The Pitcher Plant (Periuk Kera) is indigenous to Madagascar and the Indo-Malesian region, with many species concentrated around Borneo.

52 *Nepenthes fusca* Danser

THIS pitcher plant is endemic to the mossy forests of Sabah where it grows in sunny habitats on moderately high ridges and mountains. The plant is a stout climbing vine, reaching heights of up to 10 metres, with thick stalked leaves and short, erect inflorescences composed of numerous unisexual flowers. The lower pitchers are tubular in contrast to the funnel-shaped upper pitchers, but both are of similar size—about 15 centimetres long. The upper pitchers are coloured light green with varying degrees of red blotches or stripes, while the lower ones (shown here) possess blackish-purple patches or stripes. The epithet *fusca* derives from the Latin *fuscus*, 'dark', in reference to the lower pitchers. The ribbed rim (peristome) of the pitcher, a feature common to all pitcher plants, is elevated into a neck behind, and supports the narrowly triangular blackish-purple lid, unique among the Malaysian species of *Nepenthes*.

The name *Nepenthes* is derived from a Greek word meaning 'without care', and is believed to allude to the botanist's astonishment—and his forgetting past ills—upon seeing the plant.

Painted from a specimen in the Montane Garden, Kinabalu Park, Sabah.

53 *Nepenthes ampullaria* Jack

A distinctive species distributed from Peninsular Malaysia eastward to Borneo and Papua New Guinea, *Nepenthes ampullaria* is common at low elevations, particularly in swampy areas and roadside ditches. The plant is easily characterized by the presence of short round pitchers (hence the epithet *ampullaria* from the Latin *ampulla*, 'flask') produced in large masses on the ground, or in compact rosettes along the scrambling stem. The pitchers, coloured yellow-green to green with dark red spots, possess a narrow reflexed lid and a broad flattened peristome which is inclined inward. In contrast to the large leaves which develop from the high climbing stems, the leaves which produce the rosette pitchers are typically very small or nearly absent. The roots and stems of this species have been used in traditional medicine to treat dysentery and remittent fever.

Nepenthes ampullaria has been known to hybridize with another common lowland species, *N. gracilis* Korth., in areas where they occur naturally together.

Painted from a specimen near the summit of Gunung Pulai, Johor.

54 *Nepenthes rafflesiana* Jack

NAMED in honour of Sir Stamford Raffles, the founder of Singapore, *Nepenthes rafflesiana* produces some of the largest pitchers found among Malaysian species of the genus. The robust urn-shaped lower pitchers, thickly blotched red, can attain dimensions of up to 25 centimetres high and 10 centimetres across. The slender upper pitchers, measuring up to 35 centimetres high, are trumpet-shaped and coloured greenish-yellow and mottled with red. The peristome is conspicuously striped and tapers upward to support the arching orbicular lid. From densely flowered inflorescences bearing unisexual flowers, arise short woody capsules which split open, releasing the tiny winged seeds.

These climbing vines grow in wet habitats and swamps and along the margins of forests and roads, in both the lowlands and the hills. The plant's geographical range extends from Peninsular Malaysia to Borneo and Sumatra. *Nepenthes rafflesiana* was introduced into Europe last century and is the parent of many horticultural hybrids.

Painted from a specimen on Gunung Pulai, Johor.

55 *Nepenthes sanguinea* Lindley

ONE of three species of pitcher plants endemic to the mountains of Peninsular Malaysia, *Nepenthes sanguinea* grows naturally on exposed slopes and road embankments at medium elevations. The plant is a stout vine, growing up to 7 metres long, producing thick sessile leaves with bases that embrace the stem. The large cylindrical to funnel-shaped pitchers, coloured bright red to reddish-green, distinguish this species from all others which inhabit the mountains. The peristome is typically broad and flattened.

Plants of this species are said to have been collected from the peninsula and sent to England as early as the middle of the nineteenth century, where their cultivation did not meet with any great success. *Nepenthes sanguinea* is believed to hybridize naturally with the other two endemic montane species, *N. macfarlanei* Hemsley and *N. gracillima* Ridley, in areas where their distributions overlap. The epithet *sanguinea*, from the Latin *sanguis*, 'blood', alludes to the colour of the pitchers.

Painted from a specimen along the old Genting Highlands Road, Selangor.

Ferns and Fern Allies

COLLECTIVELY known as pteridophytes, the ferns and their allies are distinguished from other vascular land plants in reproducing by spores rather than by flowers and seeds. The diversity seen in their growth habits, from small aquatic plants to graceful epiphytes and large tree-ferns, is perhaps only exceeded by their numbers—over 500 species are reported from Malaysia alone. The forms we see today are descendants of a once magnificent pteridophyte flora which dominated the vegetation of the earth over 300 million years ago.

56 *Lecanopteris carnosa* (Reinw.) Blume

This epiphytic fern produces fleshy rhizomes which are hollow and inhabited by ants. The rhizomes are green when young, but become black and crusty when old, enclosing the entire branch to which they are attached. Fronds, growing up to 30 centimetres long, arise from swollen protuberances along the rhizome. The protuberances are the basis for the epithet *carnosa*, derived from the Greek *karnon*, 'a Gallic horn', alluding to their shape. The fertile lobes of the frond bear sori in cup-like depressions on marginal flaps which are folded back over the top surface of the lobe. When mature, the sporangia dehisce, releasing fine golden yellow spores. The walls of the sporangia are said to provide food to the ants, while the rhizomes provide shelter. In return, ants provide nutrition for the plant in their excreta and in the minerals they carry in from outside.

This species, distributed naturally from Sumatra to the Philippines, does show some variation, and plants confined to the high mountains of Peninsular Malaysia are assigned to the variety *pumila* (Blume) Alderw. The genus *Lecanopteris* is a member of the *POLYPODIACEAE*, a large cosmopolitan fern family well-represented in the Malaysian flora by a variety of epiphytic and terrestrial plants.

Painted from a specimen at the edge of the forest on Gunung Brinchang, Cameron Highlands, Pahang.

57 *Lycopodium complanatum* L.

Clubmoss

THE Clubmosses number about 250 species and are found from the Arctic Circle to the tropical regions of the world. *Lycopodium complanatum* is one of about 12 species of the genus native to Malaysia and occurs in the mountains, most frequently on roadcuts and among the regenerated vegetation of landslides. It is a terrestrial species with a long-creeping branching stem and tiny leaves produced in four ranks on flattened branchlets, hence the epithet *complanatum* from the Latin *complanatus*, 'flattened'. At the ends of special erect shoots are groups of cone-like strobili comprised of overlapping sporophylls, each bearing a spore-producing sporangium. In Europe, the fine spores of this species and several others were once employed as an absorbent, and as a dusting powder for chafed skin. A related species, *Lycopodium cernuum* L., has been used medicinally in Malaysia; its spores are said to contain nicotine.

The Clubmoss is classified under the *LYCOPODIACEAE*, a primitive family whose ancestors were among the earliest vascular plants to colonize the land more than 400 million years ago. The name *Lycopodium* is derived from the Greek *lykos*, 'wolf', and *podion*, 'foot', in reference to the appearance of the ends of the leafy stems.

Painted from a specimen along the roadside on Gunung Brinchang, Cameron Highlands, Pahang.

58 *Dicranopteris linearis* (Burm. f.) Underw.

Resam

THE common Resam is a characteristic feature of the Malaysian landscape, forming impenetrable thickets at the edge of forests and roads and in open places, everywhere. This is due to the long, creeping, widely spreading rhizome which regularly sends up erect stems (stipes) which bear the scrambling fronds. The frond consists of a repeatedly forked branch system, or rachis, which is largely leafless except for the ultimate branches which terminate in a pair of lanceolate leaves, and a pair of leafy accessory branches situated at each fork. The narrow lobes of the leaves bear 12 or more circular sori on their lower surface, each containing 10–15 sporangia. Long ago, the stout stems of Resam were split and used as pens for writing Arabic. They were also used in weaving to make mats, baskets, rope, and fish traps.

This species has been divided into a number of varieties which differ in the details of the branching pattern and leaflets. It belongs to the *GLEICHENIACEAE*, a primitive family of tropical and subtropical ferns with a fossil record that begins over 200 million years ago. The name *Dicranopteris* derives from the Greek *dikranon*, 'pitchfork', and *pteris*, 'fern', alluding to the forked nature of the fronds.

Painted from a specimen in the Kelang Valley, Selangor.

Select Bibliography

Burkill, Isaac H., *A Dictionary of the Economic Products of the Malay Peninsula*, 2nd edn., 2 vols., Kuala Lumpur: Ministry of Agriculture and Cooperatives, 1966. [This highly useful work is a compendium of plant, animal, and mineral products which have local socio-economic importance, arranged alphabetically by scientific or English name. Ethnobotanical aspects of over 1,200 plant species are covered. First published in 1935.]

Corner, Edred J. H., *Wayside Trees of Malaya*, 3rd edn., 2 vols., Kuala Lumpur: Malayan Nature Society, 1988. [After a general introduction to descriptive botany and the vegetation of Peninsular Malaysia, the common wayside plants are described in non-technical terms. Over 80 families of seed plants are treated, with accompanying line drawings, and over 230 black-and-white photographs. First published in 1940.]

Henderson, Murray R., *Malayan Wild Flowers*, 2 vols., Kuala Lumpur: Malayan Nature Society, 1974. [This popular work treats commonly seen shrubby and herbaceous plants of forest and wayside places. The volume on Dicotyledons was originally published as a series of three articles which appeared in *The Malayan Nature Journal* from 1949 to 1951. The companion volume on Monocotyledons was first published in 1954.]

Holttum, Richard E., *A Revised Flora of Malaya*, 3rd edn., 2 vols., Singapore: Government Printing Office, 1964. [Volume I, *Orchids*, first published in 1953, is a detailed treatment of West Malaysian orchids. The taxonomic accounts of genera and species are preceded by a general discussion of orchid morphology, classification, and cultivation. Volume II, *Ferns* (2nd edn., 1966), covers the true fern taxa, and follows the same format as Volume I.]

Keng, Hsuan, *Orders and Families of Malayan Seed Plants*, 3rd edn., Singapore: Singapore University Press, 1983. [This amply illustrated reference manual provides the diagnostic features of the orders and families of seed plants represented in Peninsular Malaysia and Singapore. Keys to the families and common genera are included. The Introduction includes a discussion of the affinities of the flora of the peninsula.]

Ridley, Henry, *The Flora of the Malay Peninsula*, 5 vols., London: L. Reeve & Co., 1922–5. [A monumental work treating all seed plants known at the time of its publication over 70 years ago. Families (referred to as Orders in the book), genera, and species are described in botanical terms, with accompanying taxonomic keys. Sparsely illustrated.]

Whitmore, Timothy C., *Tropical Rain Forests of the Far East*, 2nd edn., Oxford: Oxford University Press, 1984. [An interesting and scholarly introduction to Malesia's rain forests, emphasizing climate and seasonality, growth dynamics, forest formations, and the impact of Man. The text is well-illustrated, and includes an extensive bibliography. The first edition was published in 1975.]

Whitmore, Timothy C. and Ng, Francis S. P. (eds.), *Tree Flora of Malaya*, 4 vols., Kuala Lumpur: Longman, 1972–89. [A comprehensive treatment of the trees of Peninsular Malaysia, covering 100 families of angiosperms and gymnosperms. This authoritative account includes technical botanical descriptions, diagnostic identification keys, and line drawings. Suitable for use by professionals, students, and the interested public.]

Index to Scientific and Common Names